1. The first B-1 prototype to be part-modified to B-1B standard for flight handling tests was aircraft No. 2, seen here taking off in 1983. The B-1B is powered by four General Electric F101-GE-102 turbofan engines. It possesses no defensive armament but has the ability to carry 22 cruise missiles, or a greater number of SRAMs or other nuclear weapons, or up to one hundred and twenty-eight 500lb (227kg) high-explosive conventional bombs or fewer larger bombs.

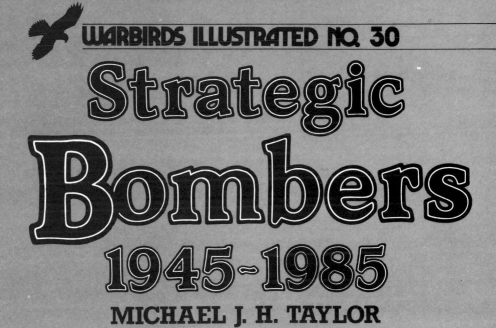

Strategic Bombers
1945~1985

MICHAEL J. H. TAYLOR

ARMS AND ARMOUR PRESS
London – Melbourne – Harrisburg, Pa. – Cape Town

Introduction

Warbirds Illustrated 30: Strategic Bombers, 1945–1985
Published in 1984 by Arms and Armour Press, Lionel Leventhal Limited, 2–6 Hampstead High Street, London NW3 1QQ; 11 Munro Street, Port Melbourne 3207, Australia; Sanso Centre, Adderley Street, P.O. Box 94, Cape Town 8000, South Africa; Cameron and Kelker Streets, P.O. Box 1831, Harrisburg, Pennsylvania 17108, USA

British Library Cataloguing in Publication Data:
Taylor, Michael J. H.
Strategic bombers, 1945–1985. – (Warbirds Illustrated; 30)
1. Bombers. – History – Pictorial works
I. Title   II. Series
623.74´63´0904   UG1242.B6
ISBN 0-85368-664-5

Editing and layout by Roger Chesneau.
Typeset by CCC, printed and bound in Great Britain by William Clowes Limited, Beccles and London.

Germany was the only nation to deploy jet bombers during the Second World War, the high performance of such aircraft being well demonstrated during March 1945 when Arado Ar 234s were hurled into action in an attempt to prevent the Allied crossing of the Rhine. Yet it was piston-engined bombers that brought the Second World War to a speedy conclusion in 1945, by means of the USAAF's devastating combination of the proven Boeing B-29 Superfortress and the atomic bomb.

After the war, the long-term future evidently lay with jet bombers, but early production jet engines were insufficiently powerful for these largest of warplanes and so piston-engined strategic bombers were initially mass-produced alongside smaller tactical jet bombers. The operational use of bombers also changed: the separate development of specialized, small, high-speed, single-seat jet fighter-bombers, themselves capable of carrying nuclear weapons, made large bombers redundant in many air forces, except for heavy raids (especially conventional) or when long range – which itself necessitated a large fuel-carrying airframe – was an operational requirement.

By the end of 1947, the race for strategic superiority between the USA and Soviet Union was under way, both nations possessing the ability to manufacture atomic weapons and both having capable bombers flying. Moreover, in that year Britain passed many examples of Rolls-Royce Nene and Derwent turbojet engines to the Soviet Union, so enabling that nation to study, develop and mass-produce modern powerplants for new fighters and tactical bombers without the need for years of research and development. In the field of rocketry, the Soviet Union was able to make considerable advances owing to the fact that the secret German ballistic missile establishment at Peenemünde was captured by Soviet forces during the final stages of the war.

The first pure jet bomber of US design was the Douglas XB-43, first flown in May 1946. This remained a prototype, and the later four-engined North American B-45 Tornado became the USAF's first jet bomber to see squadron service. However, unlike the later Boeing B-47 Stratojet, the Tornado was not strategically important. The Soviet Union's first operational jet bombers were the Tupolev Tu-12, Tu-14 and Ilyushin Il-28, all non-strategic; Britain's contemporary was the English Electric Canberra. The development in Britain of no fewer than three types of four-jet aircraft in the 'V-bomber' series finally gave the Royal Air Force strategic jet bomber capability, by which time the USA and the Soviet Union possessed intercontinental strategic jet or turboprop capability by virtue of the former's huge Boeing B-52 Stratofortress and the latter's Tupolev Tu-95 and Myasishchev M-4.

Michael J. F. Taylor

◀2
2. Boeing B-47 Stratojets became a familiar sight in many parts of the world. Here, aircraft from the USAF's 320th Bomb Wing are nearing the end of their non-stop flight from England to March AFB.

▲3 ▼4

3. The standard USAAF (United States Army Air Force) heavy bomber at the end of the Second World War was the Boeing B-29 Superfortress, the prototype of which had first flown in 1942. The initial (and major) production version was designated simply B-29, each pressurized bomber being powered by four Wright R-3350 radial engines. B-29s first went into action against Japanese targets in mid-1944 and production continued until May 1946; postwar, modernized B-29s were used to equip USAF (United States Air Force) units formed under an expansion programme. Here, Boeing Wichita-built B-29s of the US Far East Air Forces Bomber Command attack North Korean targets in early 1951. (USAF)

4. From 1945, many uses were found for B-29s other than operational deployment as bombers. Several undertook special projects, often acting as 'motherplanes' for air-launching experimental aircraft: for example, the Bell X-1, the world's first supersonic aircraft, was launched from a B-29, the first occasion being on 19 January 1946. Here a B-29 motherplane has been hoisted hydraulically so that the X-1A can be positioned under the fuselage.

5. In 1951 the US Department of Defense released this photograph of a Soviet Tupolev Tu-4, which was virtually a copy of the Boeing B-29. It was powered by four ASh-73TK engines, themselves developed from US Wrights, and could probably attain a speed comparable to the B-29's 358mph (576km/h). Tu-4s were first seen in public over the Tushino aerodrome in 1947, the type having entered production the previous year. A very small number of the Tu-4s delivered to China in the 1950s were reportedly still serviceable in the late 1970s. (USAF)

5▼

6. A larger development of the British wartime Lancaster was the Avro Lincoln, which became the RAF's last heavy bomber powered by piston engine (four Rolls-Royce Merlins). Two versions were built as new and a further variant by conversion, a B.2 being illustrated. Most Lincolns carried six gun or cannon in nose, dorsal and tail turrets, those in the nose remotely controlled from the bomb-aimer' position, and a typical bomb load was 14,000lb (6,350kg).

7. The Lincoln was also built at the Government factories in Australia for the RAAF, as the 310mph (499km/h) B.50. The first five were assembled mostly from British-built components, but later aircraft were almost completely of local manufacture. The first Lincoln B.50 flew on 17 March 1946. RAF and RAAF Lincolns were flown operationally during the troubles in Malaya. (RAAF)

8. The largest bomber ever put into operational service and the world's first strategic bomber with truly intercontinental range was the Convair B-36, conceived during the Second World War to enable USAAF squadrons to attack German targets should the whole of Europe be overrun. The XB-36 prototype (illustrated) first flew on 8 August 1946; it was powered by six Pratt & Whitney R-4360 engines carried in the 230ft (70.10m)-span wings and driving pusher propellers. The XB-36 was originally provided with single main undercarriage wheels.

9. Multi-wheel main undercarriage units were later fitted to the XB-36. This model, the subject of the photograph, was followed by YB-36 pre-production aircraft, introducing a redesigned crew compartment with the canopy raised above the fuselage top line. The initial production model was the unarmed B-36A, for crew familiarization.

10. The first fully armed B-36 was the -B variant of 1948, defensively carrying twelve cannon in six retractable, remotely controlled turrets plus two in the nose and two in radar controlled tail turrets. This B-36B is one of the aircraft subsequently modified to incorporate auxiliary jet power, becoming RB-36D long-range strategic reconnaissance aircraft or B-36D bombers.

8▲

9▲   10▼

▼11 ▲12

11. The third production version of the B-36 was the B-36D, fitted as standard with four General Electric J47 turbojet engines in pairs to improve take-off performance and increase over-target speed to 439mph (706km/h). Range was typically 7,500 miles (12,070km) and bomb load could be two massive 42,000lb (19,048kg) bombs or numerous smaller weapons.

12. A close-up photograph of the General Electric J47-GE-19 turbojet engines in pods under the wing of a B-36 in the process of construction.

13. Several other bomber and strategic reconnaissance versions followed the B-36D, including the RB-36D (illustrated) with fourteen cameras in the forward bomb bay; this variant was first flown in 1949.

▲14

14. The very long range of the B-36 led to experiments to improve the aircraft's survivability over hostile territory. As no fighter could escort the B-36, an incredibly small 'parasite' fighter (the XF-85 Goblin) was developed by McDonnell for carriage in the bomb bay, release and retrieval being achieved by means of a hook-on trapeze. Although flown, the Goblin did not become operational. However, one B-36 squadron flew modified Republic GRF-84F Thunderflash reconnaissance aircraft from specially adapted GRB-36s to undertake high-speed strategic

reconnaissance flights over selected areas. Here a Republic single-seater prepares to be retrieved by a GRB-36 motherplane.

15. On what was perhaps one of the most unusual transport operations ever, a B-36 carried the airframe of a Convair B-58 supersonic bomber from Fort Worth, Texas, to the Wright Patterson base, Dayton, for static tests in 1957.

16. A spectacular view of the gigantic B-36, an aircraft finally withdrawn from Strategic Air Command units in 1959.

▼15 16▶

▲17

17. Boeing's last piston-engined bomber was the B-50. Similar configuration to the earlier B-29, it was nevertheless a much refined aircraft: its wings, fabricated from new materials, were stronger and more efficient yet lighter, whilst other changes included a taller vertical tail and more powerful Pratt & Whitney R-4360 engines. Maximum speed of the B-50A initial production version (illustrated) was 385mph (620km/h). Deliveries to Strategic Air Command began in 1947.

18. Subsequent production versions of the B-50 medium bomber included the B-50B, examples of which are seen here with RAAF Canberra jet bombers at Jackson Field, Port Moresby, New Guinea, during exercises.

▼18

19. The principal production version of the B-50 was the B-50D, first flown in 1949 and capable of carrying an increased bomb load and extra fuel for a range of 4,900 miles (7,886km). This B-50D was subsequently modified into the single DB-50D, in order to flight-test the Bell GAM-63 Rascal stand-off missile.

20. Between 26 February and 2 March 1949 the Boeing B-50A 'Lucky Lady II', piloted by Captain James Gallagher, completed the first-ever non-stop flight round the world. Here 'Lucky Lady II' (lower aircraft) takes on fuel from a KB-29M tanker during one of four in-flight refuellings over the 94 hour 1 minute journey.

▲21

▲22

21. Out with the old and in with the new. This was the first photograph to be released showing Boeing's prototype XB-47 Stratojet, the world's first jet bomber with fully swept wings and tail surfaces, seen here nose to nose with a B-50. The XB-47 first flew on 17 December 1947. (USAF)

22. Designated a medium bomber, the B-47 followed the North American B-45 Tornado into service as the USAF's second type of jet bomber. The B-47 featured six General Electric J47 turbojet engines carried under the sweptback wings, an unusual undercarriage with tandem mainwheel units that retracted into the fuselage and outrigger units that retracted into the inboard engine nacelles, a 'bubble' cockpit canopy for the tandem-seated pilot and co-pilot (the bombardier was installed in nose), and

JATO (jet-assisted take-off) units initially built into the fuselage sides. The first major production version was the B-47B, the example here using a braking parachute.

23. B-47Bs of the 15th Air Force, SAC, on the apron at March AFB in California. Production of this model totalled 399 aircraft, of which most later underwent modification to B-47E standard and as such were redesignated B-47B-IIs. The B-47B's tail armament of remotely controlled 0.50-calibre guns was replaced by radar directed cannon in the B-47B-II.

24. The crew of a B-47B that had completed a 1,000-hour test programme, which included ninety-two bombing runs (twelve dropping live weapons), in early 1953. The crew sit on a pile of dummy bombs!

23▲ 24▼

17

12422

UNITED STATES AIR FORCE

25. The use of Aerojet JATO rockets greatly increased the B-47's thrust at take off.

26. The sting in the B-47E's tail: two 20mm cannon in their General Electric turret, the latter controlled by a computer.

▲27 ▼28

7. The major production version of the B-47 was the B-47E, first flown in January 1953. This had more powerful J47 engines, was armed with two cannon in the tail and could carry 20,000lb (9,070kg) of bombs, and many were later modified to B-47E-II standard. Speed and range were 606mph (975km/h) and 4,000 miles (6,437km) respectively. B-47 production ended in February 1957, and aircraft were gradually transferred from Strategic Air Command to Tactical Air Command as new types of bomber appeared. These B-47Es are about to take part in a night exercise.
8. This ceremony marked the production of the 1,000th B-47, an E variant destined for the 40th Medium Bombardment Wing (a unit of the 15th Air Force) at Smoky Hill AFB, Salina, Kansas.
9. A long-range photographic reconnaissance version of the B-

47E, for use by day or night, was the RB-47E, which incorporated a compartment for eleven cameras and photoflash bombs. This RB-47E is taking on fuel from a KC-97 Stratotanker using the Flying Boom technique.
30. Many specialized variants of the B-47 were produced by modification of earlier bombers, roles including weather reconnaissance, electronic countermeasures, crew training and operation as pilotless drones to test defences. A few B-47Bs and -Es were modified into YDB-47Bs and YDB-47Es, experimental test aircraft for the Bell GAM-63 Rascal stand-off missile. A YDB-47E is illustrated, its missile carried on a rack mounted on the starboard fuselage side.

▲31 ▼32

33▲

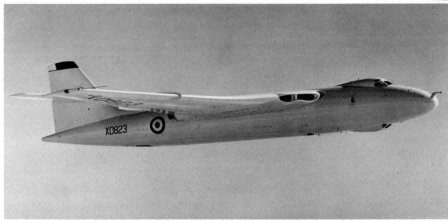

34▲

31. The first of Britain's 'V-bombers' to see RAF service, and the first to be phased out (prematurely, because of wing structure fatigue, in 1964), was the Vickers Valiant. It had first flown on 18 May 1951 and was Britain's first four-jet bomber. Illustrated is the first prototype.

32. The initial production version of the Valiant was the B.1, which entered Bomber Command service (replacing Avro Lincolns) from 1955 and was used during the Suez Crisis of 1956. Powered by four Rolls-Royce Avon turbojet engines carried within the wings, it could attain a speed of Mach 0.84 and had a range of 4,500 miles (7,240km). The bomb load was up to 21,000lb (9,525kg) and no defensive armament was carried. The later Valiant B(PR)K.1 was a bomber and photographic reconnaissance version, whilst the capability of the aircraft was further increased in the B(PR)K.1, suited to bombing (with in-flight refuelling capability), photographic reconnaissance and in-flight refuelling tanker duties (as illustrated). (Ministry of Defence)

33. The probe on the nose of this Valiant B(PR)K.1 manoeuvring to receive fuel is just visible, as are the auxiliary underwing fuel tanks (to extend range) and the glazed fairing of the bomb-aimer's position. (Ministry of Defence)

34. The final Valiant model was the B(K).1, a bomber and in-flight refuelling tanker (receiver) able to transfer about half its maximum fuel load of 9,972 gallons (45,330l). This B(K).1 was the first Valiant to be flown in an all white anti-radiation finish.

▲35

▲36 ▼37

35. On 11 October 195[ ] a Valiant of No 49 Squadron, RAF, dropped the first atom[ ] bomb to be released from a British aircraft, over Southern Australia, whilst an aircraft of the same squadron dropped Britain's first hydroge[n] bomb near Christmas Island on 15 May 1957[ ] Here a Valiant B(K).1 refuels a Vulcan. (Ministry of Defence[)]

36. The huge, eight-engined Boeing B-52 Stratofortress remains fully operational today in reduced numbers as the USAF's and NATO's only heavy strategic bomber, a situation that will continue until the Rockwell Internationa[l] B-1B goes into service in the late 1980s; even then, the B-52 will continue to play an important first-line rol[e] with the USAF. Two prototypes were built, the XB-52 and the YB[-]52, first flying on 2 October and 15 April 1952 respectively. Each aircraft was powered b[y] eight Pratt & Whitney J57 turbojet engines an[d] featured a B-47 style cockpit with the pilot and co-pilot seated in tandem. Here, the XB[-]52 makes a taxi run in crabbed position in hig[h] cross-winds.

37. The first production model of th[e] Stratofortress was the B-52A, although only three such aircraft wer[e] completed. Featuring [a] new-style stepped cockpit with side-by-side seating for the pilo[t] and co-pilot, the first example was flown on [ ]August 1954.

38. A Boeing B-29A Superfortress with a four-gun forward dors[al] turret.

39. B-29s were armed with guns in four remotely controlled, electrically operated fuselage turrets and [a] tail turret, plus up to 20,000lb (9,072kg) of bombs. The forward pressurized compartment accommodated the pilot, co-pilot, bombardier, navigator[,] radio operator and engineer.

▲40 ▼41

40. Close-up of the tail
gun position of a B-29A.
41. The prototype
Convair B-36,
recognizable by its flush
cockpit glazing.
42. An update for the
B-52G was the
replacement of Hound
Dogs with twenty
Short-Range Attack
Missiles (SRAMs) and
free-fall bombs.
However, the most
recent armament of the
B-52G is ALCMs (Air-
Launched Cruise
Missiles) in the form of
AGM-86Bs, with fewer
SRAMs and bombs.
The aircraft illustrated
is a SRAM-carrying B-
52G, with six missiles
under each wing and
eight on a rotary
launcher in the bomb
bay.
43. An impressive
display of Boeing B-47
Stratojets.

▲44 ▼45

44. This close-up view of a Boeing Stratofortress shows how cruise missile-carrying B-52Gs have been given wing root 'strakelet' fairings (rounding off wings with fuselage) for easy identification, in accordance with the unratified SALT-II strategic arms limitation talks.

45. The B-52H is another carrier of the ALCM and, like the B-52G, was originally modified as a SRAM bomber. Each B-52G in the latest ALCM configuration carries six such weapons under each wing pylon.

46. The B-52B was the first major production version of the Stratofortress, the fifty built including RB-52B aircraft capable of undertaking bombing, photographic reconnaissance and electronic countermeasures roles. Here one of three B-52Bs to complete the first non-stop round-the-world jet flights (refuelled by KC-97s) lands at March AFB with its drag parachute billowing. It had achieved this record in a flying time of 45 hours 19 minutes, between 16 and 18 January 1957. All three had taken off from Castle AFB, home of the Strategic Air Command's 93rd Bomb Wing.

47. The B-52C was a development of the RB-52B, but the B-52D was intended only as a long-range heavy bomber. Up to the -D variant, B-52 production had been undertaken exclusively at Seattle; the B-52D was the first Wichita-built Stratofortress, as announced on the fuselage of the aircraft illustrated here. The first B-52D flew on 14 May 1956, and the variant accounted for 170 of the 744 Stratofortresses built. Armament was four 0.50-calibre machine guns in a manned tail turret, plus up to 60,000lb (27,215kg) of bombs carried in the bay and underwing.

47▼

▲48

48. To improve upon the B-52D's bombing, navigation and electronic systems, Boeing produced the B-52E, which appeared in October 1957. This B-52E is taking on fuel from a Stratotanker.
49. More powerful versions of the J57 engine had been fitted progressively to B-52s as production continued through to the B-52F. The B-52G, however, marked a major update in design: the tail fin, rudder and wings (the latter to carry integral fuel tanks) were redesigned, and the tail gunner's position was deleted and replaced by a new turret which could be operated remotely by the

gunner now seated in the forward pressure cabin and using closed-circuit television or by a new automatic fire control system. The B-52G first entered service with the USAF in early 1959.
50. The B-52G was developed to carry two AGM-28 Hound Dog supersonic missiles under its wings (as seen here) as well as Quail decoy missiles in the bay. Representing the major production version of the B-52, all 193 had been delivered by early 1961. Speed and unrefuelled range are 595mph (957km/h) and 7,500 miles (12,070km) respectively. (USAF)

▼49

51. On two days in early August 1964, US Navy vessels were attacked by North Vietnamese patrol boats; in retaliation, Navy aircraft struck at North Vietnamese naval bases the day after the second attack, marking the beginning of US air operations over Vietnam. A major concern of the latter was the attempt to halt the Viet Cong in South Vietnam by bombing supply routes and bases. Here three B-52s set out on such a mission in October 1966. (USAF)

52. Together with the B-52H, the B-52G is currently one of the major operational versions of the Stratofortress in USAF service, and over the years it has been the subject of many improvement programmes devised to extend its capabilities and useful life. This B-52G, for example, has two bulges under its nose, indicating the addition of an Electro-optical Viewing System (EVS) to enhance its low-level penetration capabilities: one houses a Forward-Looking Infra-Red (FLIR) scanner and the other a low-light television (LLTV) camera.

53. The final production version of the Stratofortress was the B-52H, which appeared in 1961. The only version to switch from J57 turbojets to eight Pratt & Whitney TF33-P-3 turbofan engines, and with a single multi-barrel cannon as tail armament, it has a maximum speed similar to that of the -G but a range of 10,000 miles (16,093km) with in-flight refuelling. It was developed originally to carry four Skybolt air-launched ballistic missiles but these were not adopted for service. Here a B-52H poses with Skybolts during development trials.

▲51

52▲   53▼

33

▲54

54. The prototype of the Soviet Tupolev Tu-16 flew for the first time in 1952 and full production began the following year. It was designed as an intermediate-range medium bomber powered by two Mikulin AM-3 turbojet engines partly recessed in the fuselage sides, and it is thought that about 2,000 were built, of which a large number still remain in squadron use in various roles. The initial strategic bomber version was given the name 'Badger-A' by NATO. This 616mph (992km/h) aircraft has seven guns in dorsal, ventral, tail and nose positions and carries up to 19,840lb (9,000kg) of bombs. Some 'Badger-As' are now used as in-flight refuelling tankers; here, an example of this version refuels a bomber. (Ministry of Defence)

55. 'Badger-B' is the NATO name for the version of the Tu-16 originally armed with two 'Kennel' air-to-surface missiles. The aircraft illustrated, with 'Kennels' under its wings, was one formerly operated by the Indonesian Air Force. Today, 'Badger-Bs' are still flown by the Soviet Dalnaya Aviatsiya, but they are armed with gravity bombs.

56. 'Badgers' are flown also by Soviet Naval Aviation and these aircraft, plus some Air Force bombers, can undertake anti-shipping roles. 'Badger-C' can be armed with an underfuselage 'Kipper' missile, the attachment point for which is clearly visible in the photograph, or new 'Kingfish' missiles under the wings (one shown). Late-build 'Badgers' were fitted with more powerful RD-3M engines. (Royal Danish Air Force)

57. Many of the better photographs of Tu-16s have been taken while the aircraft are flying close to NATO countries or near NATO ships on exercise. This 'Badger', its bulged nose radome (in place of a glazed nose with gun) indicating it to be either a 'Badger-C' or a maritime and electronic reconnaissance 'Badger D', was photographed by US Navy fighters as it flew high over USS *Kitty Hawk* on exercise in the Pacific Ocean in 1963. Clearly to be seen are the tail guns, the dorsal twin-gun barbette and the housings projecting rearwards from the wings into which the main undercarriage units retract. (US Navy)

▼55

▲58 ▼59

58. A 'Badger' flying close to a British aircraft carrier on exercise, the three ventral blisters under the centre of the fuselage identifying it a -D variant. This aircraft also shows the ventral barbette and guns. (Ministry of Defence)

59. A striking photograph of the rear guns and tail pressure cabin area for the gunner and radio operator of a Tu-16, taken by the crew of a Royal Navy Phantom from HMS *Ark Royal* in 1970. (Ministry of Defence)

60. 'Badger-G' was put into service with Soviet Naval Aviation to carry two 'Kelt' missiles or free-fall bombs; others were delivered to Egypt. Some, like the aircraft illustrated, have been modified to launch two 'Kingfish' missiles with either nuclear or high explosive warheads.

61. 'Badger-E' is a photographic reconnaissance aircraft whilst 'Badger-F' is similar but can be distinguished by pods housing electronic intelligence equipment carried on pylons under the wings. During maritime flights, 'Badger-Fs' are often accompanied by a different variant of the same aircraft. Here, a US Navy F-4 Phantom and an F-8 Crusader from USS *Kitty Hawk* escort a 'Badger F' (forward aircraft) and another 'Badger' flying in the vicinity of the carrier. (US Navy)

62. (Next spread) Other versions of 'Badger' in Soviet use include models for electronic countermeasures and electronic reconnaissance duties. China currently operates a potent force of some 120 'Badger-A' types, built in China since the late 1960s at the Xian works as the Hong-6 (as illustrated). These have Wopen-8 engines, which are in essence locally produced RD-3Ms, and can carry nuclear or conventional weapons. (Liu Zhibin)

60▲   61▼

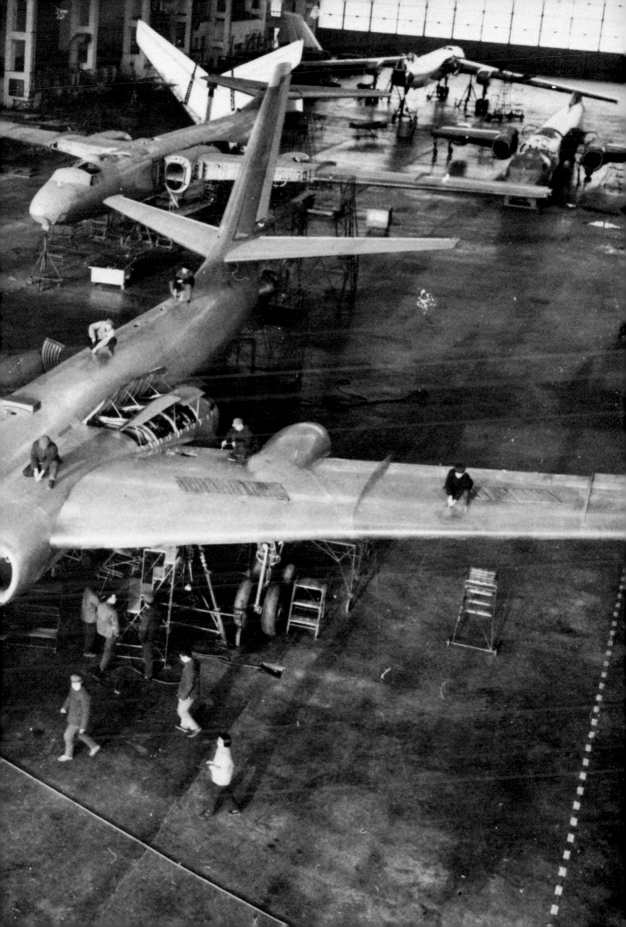

▲63

▲64 ▼65

63. From 1949 Avro flew a series of small, mostly single-seat, research aircraft built to test the behaviour of delta wings at low and high speeds. Using this data, Avro completed the design of its Vulcan long-range medium bomber, destined to be the RAF's second 'V-bomber' type in service. Two prototype Vulcans were built, the first (illustrated) flying initially on 30 August 1952.

64. A Vulcan B.1 takes on fuel from a Valiant B(PR)K.1. Later, the type of anti-radar jamming avionics installed in the Vulcan B.2 was fitted to a number of B.1s in enlarged tailcones, the modified aircraft becoming Vulcan B.1As.

65. Like the prototypes, the first few production Vulcan B.1 bombers had delta wings with straight leading-edges. However, wings with compound sweep on the outer sections became standard, to reduce buffeting effects at high altitude. Vulcan B.1s entered RAF service from August 1956, each powered by four Bristol Siddeley Olympus turbojet engines (which were progressively increased in power as production continued).

66. Until the recent withdrawal from RAF service of Vulcan bombers, the USAF's Boeing B-52 and the Vulcan were partner bombers within NATO.

66▼

▲67

67. With the availability of more powerful Olympus turbojet engines, Avro developed the Vulcan B.2. As well as increased power, the B.2 featured larger wings with elevons in place of the B.1's ailerons and elevators, and survivability was increased by the adoption of anti-radar jamming avionics in a bulged tailcone. The B.2 was intended as a carrier of Britain's Blue Steel stand-off nuclear missile or the Douglas Skybolt air-launched ballistic missile. Here an early B.2 carries Skybolts during development trials; two missiles are fitted, on specially designed underwing pylons.

▼68

8. In the event, Skybolt failed to be selected for the RAF's Vulcan or any other aircraft, leaving Blue Steel as the main weapon of the Vulcan B.2. The missile was carried semi-recessed under the fuselage. This is a B.2 of No 27 Squadron.
9. As an alternative to Blue Steel, the Vulcan (and earlier versions) could carry twenty-one 1,000lb (454kg) bombs. The B.2 had a cruising speed of 625mph (1,005km/h) and a combat radius of 2,300 miles (3,700km) at high altitude without in-flight refuelling. No defensive armament was carried.

70. As time passed, operational B.2s were modified to allow low-level penetration of the target area to evade anti-aircraft missile systems. The original white anti-flash finish was, therefore, replaced by a fully camouflaged upper surface. Some B.2s were also converted into SR.2 strategic reconnaissance aircraft.

71. A Vulcan B.2 in final configuration flown by No 230 OCU from Waddington, assigned the task of overland strike and armed usually with twenty-one 1,000lb (454kg) high-explosive bombs. It had been intended to withdraw all Vulcans from RAF service by the summer of 1982, but the conflict over the Falkland Islands put back the final phase-out until December 1982. During the South Atlantic War, Vulcans performed the longest-range bombing missions in aviation history, using in-flight refuelling to extend endurance for the 15–16-hour flights between Ascension Island and Port Stanley airfield. (Brian M Service)

72, 73. A squadron (No 83) of majestic Avro Vulcan B.2s during their heyday in RAF service.

▲70 ▼71

74. A Soviet Tupolev Tu-95 'Bear' photographed from a Convair F-102 Delta Dagger of the 57th Fighter Interceptor Squadron, USAF, based at Keflavik, Iceland, as it flew a few hundred miles east of Iceland. (USAF)

75. A Strategic Air Command General Dynamics FB-111A with SRAM attack missiles pylon-mounted under the wings.

76. Easily the most formidable bomber in operational service today is the Soviet Tupolev Tu-22M, known to NATO as 'Backfire'. A medium bomber with swing wings and powered by turbofan engines, it has a weapon load of about 26,450lb (12,000kg), which is light when compared to that of the FB-111A or B-52, but this is offset by its ability to fly at twice the speed of sound at high altitude or near Mach 1 at low level and carry up to

three 'Kitchen' or 'Kingfish' air-to-surface missiles or nuclear or conventional bombs. It will certainly also be a carrier of new Soviet cruise missiles. Advanced electronic systems, the deployment of decoy missiles to confuse enemy defences and twin radar-directed tail guns all help to increase the aircraft's survivability. (USAF)

77. The latest French Air Force single-seat interceptor fighter is the Mach 2.3 Dassault-Breguet Mirage 2000; from this has been developed the Mirage 2000N, a two-seat, low-altitude penetration aircraft that will be a carrier of the ASMP supersonic nuclear missile. First flown in 1983 in prototype form, production Mirage 2000Ns will appear from 1986.

▲78
78. The second Rockwell International B-1 prototype, now flying as the first B-1B testbed prototype, undertakes in-flight refuelling manoeuvres.
79. The very latest Soviet strategic bomber, known to NATO as 'Blackjack', is currently under test and evaluation and is expected to enter operational service from about 1986. This is an artist's impression of the four-engined bomber, which has been attributed to Tupolev. Larger than the USAF's B-52 and future B-1B and undoubtedly capable of Mach 2, it will be able to carry cruise missiles or other nuclear or conventional weapons.'Blackjacks' will replace Tu-95s in the strategic role. (USAF)

▼79

80▲

80. The last of the three types of 'V-bomber' to enter RAF service as the Handley Page Victor. It featured a unique crescent wing, first flight-tested in scale form on the Handley Page HP.88 research aircraft. The crescent wing had originated in wartime Germany, the theory being that as the wing became less swept and had reduced thickness/chord ratio from root to tip the critical Mach number would remain the same along the whole span. The first of two Victor prototypes (illustrated) flew in late 1954.

81. Production Victor B.1s (intended, like other early 'V-bombers', to carry free-fall conventional or nuclear bombs) were ordered before the prototypes flew, each to be powered by four Bristol Siddeley Sapphire turbojet engines buried in the wings. Here Victors take shape on the final assembly lines at the Colney Street works.

81▼

82. The very first production Victor B.1s: the earliest examples were used for development trials and tests, whilst others joined No 232 Operational Conversion Unit in November 1957. The first fully operational RAF Victor squadron was No 10 at Cottesmore in early 1958. Note that the Victor B.1 lacked the dorsal fillet forward of the tail fin that was a feature of the later B.2.

83. In 1959, this early production Victor underwent trials using two Spectre rocket motors to assist take-off at high weight. These motors added 16,000lb (7,256kg) of thrust to that of the four turbojets, and take off with maximum bombload and fuel could be achieved in 1,650ft (503m).

84. The Victor carried no defensive armament, relying instead on high performance for survivability. The offensive load in the conventional role was thirty-five 1,000lb (454kg) bombs.

▲82 ▼83

84▲

85. In 1962, No 139 Squadron became the first to be equipped with the improved Victor B.2, one of its aircraft being shown here at Wittering. The main update was the use of four much more powerful Rolls-Royce Conway engines in wings of somewhat greater span, but other refinements included enlarged air intakes and the addition of a dorsal fillet ahead on the tailfin; later streamlined pods were added above the trailing edge of each wing to reduce drag, these containing chaff which could be released to confuse enemy radar. The speed and range of this version were Mach 0.92 and 4,600 miles (7,400km) respectively. Twenty-one of the 34 Victor B.2s were also equipped to carry the Blue Steel missile, as seen fitted beneath this aircraft. (Ministry of Defence)

86. Victor B.1s, when fitted with electronic countermeasures equipment in the rear fuselage, were redesignated B.1As. When Mk 1s and 1As passed out of RAF service as bombers, many were modified into K.1 and 1A tankers. The photograph shows a K.1 three-point tanker of No 57 Squadron from RAF Marham. (Ministry of Defence)

87. Nine Victor B.2s were eventually modified as strategic reconnaissance aircraft under the designation Victor SR.2, capable of radar mapping a 750,000 square mile area during a six-hour mission. These served with No 543 Squadron until Vulcans took over this role. Here, reconnaissance cameras are loaded on board an SR.2. (Ministry of Defence)

88. From 1974, Victor B.2s (already some years out of service as bombers) and SR.2s became operational again as K.2 in-flight refuelling tankers with reduced wingspans. During the Falklands conflict Victor three-point tankers flew nearly six hundred missions with virtually total success. This K.2 is refuelling a prototype Panavia Tornado.

▲85 ▼86

89. The Soviet Union's first four-turbojet strategic bomber in operational service was the huge Myasishchev M-4, which was first seen in public in 1954. Powered by four Mikulin AM-3D engines buried in the wing roots, it looked more modern than its contemporary, the Tupolev Tu-95, but in fact was less successful and was built in much smaller numbers. 'Bison-A' (NATO name) was the only bomber version, armed with ten cannon for self-defence because of its relatively low operational ceiling. Its speed was 620mph (998km/h). Only about 40 remain in Soviet squadrons as bombers, although others serve as in-flight refuelling tankers. 'Bison-B' (illustrated) is the basic armed maritime reconnaissance and anti-submarine warfare model, with six guns and a solid nose instead of the bomber's glazed nose. (US Navy)

90. An improved maritime reconnaissance and ASW version is 'Bison-C' also still operational. This model is identifiable by its longer nose, housing large search radar, and central refuelling probe. As with other versions, the -C has an undernose fairing to accommodate a bombardier/observer in a prone position.

▼89

▼90

. Despite first impressions, the Soviet Tupolev long-range
rategic and maritime reconnaissance bomber, known to NATO
'Bear', is an extremely capable first-line aircraft, and the Soviet
alnaya Aviatsiya operates more than 100 'Bear-As', -Bs and -Cs
Tu-95 strategic bombers. The Soviet Naval Aviation retains
me 75 'Bears' for maritime strike and reconnaissance duties as
u-142s. Each aircraft is powered by four Kuznetsov NK-12M
engines (the largest turboprops ever developed), driving contra-
rotating propellers. Capable of 575mph (925km/h) and possessing
a range of 7,800 miles (12,550km) with a 25,000lb (11,338kg)
bombload, 'Bear-A' carries nuclear or conventional weapons
internally; the more important 'Bear-B' (illustrated) can launch
externally attached 'Kangaroo' or 'Kitchen' air-to-surface
missiles. (US Navy)

▲92

92. Most 'Bears' carry as defensive armament twin cannon in dorsal and ventral remotely controlled barbettes and in a crewed tail position. This photograph shows a 'Bear-C', a variant generally similar to the 'Bear-B' and one which retains the ability to carry 'Kangaroo' (the largest airborne missile ever to become operational). This particular photograph was taken in 1964 during NATO exercise 'Teamwork', when the 'Bear-C' was first seen by Western observers.

93. 'Bear-D' is a naval version used for reconnaissance and for assisting missile-carrying aircraft and ships to aim and guide their weapons on to targets. This example is being escorted by an RAF Lightning from No 23 (Red Eagle) Squadron, of the Northern Interceptor Force from RAF Leuchars, which had located it flying at high altitude over the North Sea. (Ministry of Defence)

94. The 'Bear-D' carries a radar scanner in an undernose fairing and a large underfuselage radome housing X-band radar for locating enemy shipping and for reconnaissance purposes. This model was first photographed in detail when several examples flew over the US Coast Guard icebreakers *Eastwind* and *Edisto* in the Kara Sea in 1967 as they attempted to circumnavigate the Arctic. Here, a 'Bear-D' passes low over *Edisto*. (US Coast Guard)

95. The Tu-95/-142 has been so successful that limited production continued into the 1980s in order to maintain a constant operational strength. The 'Bear's range and endurance can be extended by in-flight refuelling, the nose probe of this aircraft showing clearly. Other variants are the 'Bear-E' for maritime reconnaissance and the anti-submarine 'Bear-F'. (Ministry of Defence)

▼93

96. The world's first supersonic strategic bomber was the Convair B-58 Hustler, a delta-winged three-seater powered by four General Electric J79 turbojets; speed and range were 1,385mph (2,230km/h) and 2,400 miles (3,860km) respectively. The first B-58 prototype (illustrated) made its maiden flight on 11 November 1956.

97. The first prototype Hustler takes off from Convair's Fort Worth plant, the sixteen tyres of the main undercarriage units creating a cloud of dust.

96 ▶

▼97

▲98

99▶

98. The B-58's crew of three sat in individual cockpit capsules, allowing them to survive ejection in an emergency at supersonic speeds. However, the most obvious design feature of the Hustler was its detachable underfuselage pod, used to house nuclear or conventional weapons and the aircraft's fuel for the journey to its target, thereafter being jettisoned before the return journey, for which the aircraft took fuel contained in its wing tanks.

99. The range of the Hustler could be improved by the use of in-flight refuelling, two such link-ups making eighteen-hour missions possible.

100. A view of the B-58 Hustler without its underfuselage pod.

101. Production of the B-58A Hustler totalled 86 aircraft, and the first assigned Strategic Air Command unit became operational in early 1960. Some development B-58s were also brought up to production standard.

102. Eventually a two-component pod was developed for the B-58A, the lower, larger section carrying fuel and the upper the weapon load plus fuel, electronic countermeasures devices or reconnaissance equipment.

▲103

103. The two-component pod in position under the fuselage of an early Hustler.

104. The Hustler served through the 1960s but was withdrawn from service in 1970 as an economy measure. This photograph shows a B-58A with pods and weapons and the radar-directed 20mm Vulcan multi-barrel cannon of another aircraft normally fitted as tail armament.

▼104

05. By scaling-up the general configuration of the Mirage III ghter and providing two SNECMA Atar 9 turbojets as owerplants, Dassault produced France's first supersonic strategic omber, the Mirage IV. A two-seater capable of Mach 2.2 and ossessing a tactical radius of 770 miles (1,240km), the prototype ew for the first time on 17 June 1959. Three pre-production aircraft were followed by 62 full production Mirage IVAs with Atar 9K-50 turbojets.

106. As France's nuclear deterrent, the Mirage IVA carried a free-fall nuclear weapon semi-recessed under the fuselage. A dummy weapon can be seen here in position for a test flight of the bomber.

107. A Mirage IVA refuels from one of the French Air Force's C-135F tankers.

108. When French Navy submarines and land missiles took over the main deterrent role, Air Force Mirage IVAs were assigned new duties as tactical strike aircraft. Today, thirty or so are retained for this role (with a few in reserve), each carrying tactical nuclear weapons, four Martel air-to-surface missiles or sixteen 1,000lb (454kg) bombs; four more are configured as strategic reconnaissance aircraft (Mirage IVR). From about 1985 the remaining bombers will be armed with ASMP supersonic nuclear stand-off missiles for use against well-defended targets such as military airfields. (French Air Force)

109. First seen publicly in 1961, the Tupolev Tu-22 was the Soviet Union's first supersonic bomber. Production was restricted to about 250 aircraft because its range of only about 1,925 miles (3,100km) without in-flight refuelling greatly diminished its strategic capabilities. Nevertheless, Tu-22s remain in operational use today in both bomber and maritime reconnaissance-bomber roles. The initial version, which entered service armed with free-fall bombs in a weapons bay, is known to NATO as 'Blinder-A'; 'Blinder-B' is similar but was configured to carry a 'Kitchen' air-to-surface nuclear missile semi-recessed under the fuselage. This Mach 1.4 'Blinder-B', its in-flight refuelling nose probe retracted was photographed in 1961. (US Navy)

110. Small numbers of 'Blinder-Bs' are flown by the air forces of Libya and Iraq. Here, US Navy F-4N Phantoms from VF-51 escort intercepted Tu-22s that were located flying over the Mediterranean on delivery from the Soviet Union to Libya. (US Navy)

111. 'Blinder-C' and 'Blinder-D' are maritime reconnaissance-bomber and twin-cockpit trainer versions respectively, both in current use. Despite the age of the Tu-22, many of its details remain unknown (for example, the two large turbojet engines carried one each side of the tail fin are of unknown type). This dramatic view of a Tu-22 was taken by a Royal Danish Air Force fighter.

▼107

▼108

109▲

110▲ 111▼

▲112

113▼

112. Following the successful completion of the difficult task of developing the world's first swing-wing supersonic aircraft, in the form of the F-111, it was a comparatively straightforward procedure for General Dynamics to modify the design to produce a two-seat strategic bomber derivative. This became the FB-111A, the prototype of which (illustrated) flew for the first time on 30 July 1967.

113. The FB-111A is capable of Mach 2.5 on the power of two Pratt & Whitney TF30-P-7 turbofan engines and has a range of more than 4,000 miles (6,437km) with auxiliary fuel. The production programme was cut from the original 253 aircraft to just 76, mainly as an economy measure. The first of these entered Strategic Air Command service in 1969.

▲114

◄116

114. Up to forty 750lb (340kg) bombs in clusters can be carried by the FB-111A under its wings, with a further two in the internal bay, although the number of bombs is reduced with the angle of wing sweep; alternatively, six nuclear bombs or six SRAMs can be carried, four under the wings. Here an FB-111A test-launches a SRAM, released from the bay.

115. It is expected that the Soviet forces will deploy about 400 'Backfires', of which some 250 or more may already be in use. The aircraft illustrated has a 'Kitchen' missile under its fuselage. With a combat radius of about 3,400 miles (5,470km), it is, clearly, designed primarily for operation in Europe, but some are also based in the Pacific area. The Soviet Naval Aviation also uses 'Backfires' as missile-carrying anti-shipping aircraft, which are proving much more formidable than earlier types ranging over the Atlantic: with in-flight refuelling, even the American continent is within range. (USAF)

116. Realizing that the Boeing B-52 Stratofortress needed replacing in USAF service if America was to retain a credible triad deterrent force – comprising strategic bombers, nuclear ballistic missile submarines and land-based missiles – US designers initiated studies in the late 1960s to define a new bomber. In 1970 Rockwell International was awarded the contract to develop such an aircraft as the B-1. This first flew in prototype form on 23 December 1974 and was a Mach 2 swing-wing bomber with intercontinental range. Eventually 244 production B-1s were ordered for the USAF, all of which would have been in service by 1981 had not the programme been cancelled in 1977.

▲117
117. In 1981 the B-1 programme was revived but in modified form, just 100 B-1Bs being ordered for USAF operational service from 1986. These will have the ability to fly at only about Mach 1.25 but they will incorporate the latest technologies in avionics and possess a very low radar signature. Here one of the original

prototype B-1s poses for the camera with wings fully spread for low-speed flight.
118. This side view of a B-1 prototype shows the extremely clear lines of the bomber, its lack of sharp angles being a key factor in the 'low observable' technology to reduce its radar signature.

▼118